Nelson Inte Science
Workbook 1

Anthony Russell

OXFORD
UNIVERSITY PRESS

Great Clarendon Street, Oxford, OX2 6DP, United Kingdom

Oxford University Press is a department of the University of Oxford.
It furthers the University's objective of excellence in research, scholarship,
and education by publishing worldwide. Oxford is a registered trade mark of
Oxford University Press in the UK and in certain other countries

Text © Anthony Russell 2012
Original illustrations © Oxford University Press 2014

The moral rights of the authors have been asserted

First published by Nelson Thornes Ltd in 2012
This edition published by Oxford University Press in 2014

All rights reserved. No part of this publication may be reproduced,
stored in a retrieval system, or transmitted, in any form or by any
means, without the prior permission in writing of Oxford University
Press, or as expressly permitted by law, by licence or under terms
agreed with the appropriate reprographics rights organization.
Enquiries concerning reproduction outside the scope of the above
should be sent to the Rights Department, Oxford University Press, at
the address above.

You must not circulate this work in any other form and you must
impose this same condition on any acquirer

British Library Cataloguing in Publication Data
Data available

978-1-4085-1726-0

10 9 8

Printed in India

Acknowledgements
Cover illustration: Andy Peters
Illustrations: David Benham, Moreno Chiacchiera, Simon Rumble and Wearset Ltd
Page make-up: Wearset Ltd, Boldon, Tyne and Wear

The authors and the publisher would like to thank Judith Amery for her contribution to the development of this book.

The authors and the publisher would like to thank the following for permission to reproduce material:

p.5: (cat) Fotolia/Eric Isselée, (plant) iStockphoto/Valentyn Volkov.

Although we have made every effort to trace and contact all
copyright holders before publication this has not been possible in all
cases. If notified, the publisher will rectify any errors or omissions at
the earliest opportunity.

Links to third party websites are provided by Oxford in good faith
and for information only. Oxford disclaims any responsibility for
the materials contained in any third party website referenced in
this work.

Contents

1 Plants — 3

Living things — 3
    Activity 1 — 3
    Activity 2 — 5

Things that have never lived — 7
    Activity 3 — 7

Plants and animals in their environments — 10
    Activity 4 — 10

The parts of a plant — 18
    Activity 5 — 19

What plants need to grow — 22
    Activity 6 — 22

Growing plants from seeds — 26
    Activity 7 — 26
    Activity A — 31
    Activity B — 33

2 Humans and animals — 35

We are all different – and the same — 35
    Activity 1 — 35

Body parts — 38
    Activity 2 — 38
    Activity 3 — 39
    Activity 4 — 41

Health and diet — 43
    Activity 5 — 43
    Activity 6 — 44

Senses — 46
    Activity 7 — 46
    Activity 8 — 47
    Activity 9 — 49

Parents and children — 53
    Activity 10 — 53
    Activity 11 — 55
    Activity C — 57
    Activity D — 58

Contents

3 **Material properties** 60

What are things made of? 60
   Activity 1 60

Materials and their properties 62
   Activity 2 62

Naming materials 64
   Activity 3 65

Sorting materials 67
   Activity 4 67
   Activity E 69
   Activity F 70

4 **Forces** 72

Movement 72
   Activity 1 72

Pushes and pulls 78
   Activity 2 80

Changes in movement 81
   Activity 3 81
   Activity G 83
   Activity H 84

5 **Sound** 85

Sources of sound 85
   Activity 1 85

Sound and distance 87
   Activity 2 87

Hearing sounds 89
   Activity 3 89
   Activity I 92
   Activity J 94

Introduction

Nelson International Science Workbook 1 provides a complete copy of the *Student Book* activities for all learners to work through.

The activities are marked with 📖 showing the corresponding page number in the *Student Book*.

In addition to the *Student Book* activities, there are extra activities marked, for example, Activity A, that can be done in the classroom or as homework at home. They support the knowledge and understanding gained in the *Student Book* activities.

Chapter 1: Plants

Living things

(a) grains

(b) tree

(c) fish

(d) bird

(e) leaf

(f) maize

(g) tooth

(h) fruit

(i) shell

(j) insect

(k) animal breaking out of shell

(l) bone

(m) earthworm

(n) dog

(o) flower

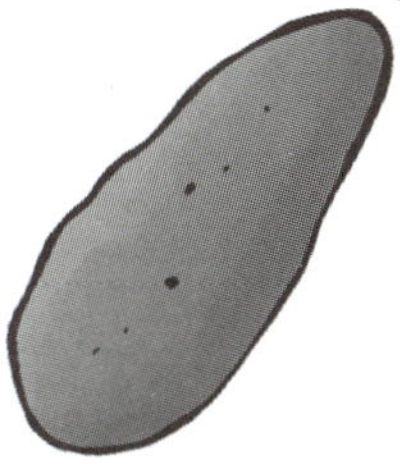

(p) sweet potato

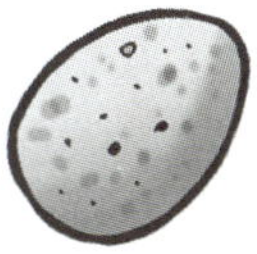

(q) bird's egg

Activity 1

You will need: a pen or pencil.

1 Look at the pictures.

Chapter 1: Plants

Activity 1 (continued) 📖

2 Sort the things into two groups — a plant group and an animal group.

3 Write down the letters of five of the things in each group (record your answers).

Find five plants:

Plant group

☐ ☐ ☐ ☐ ☐

Find five animals:

Animal group

☐ ☐ ☐ ☐ ☐

4 Share your answers (communicate) with the class.

Living things

Activity 2 　5

You will need: a pen or pencil.

🔍 **1** Look at the pictures of the animal and the plant.

2 How are they different? Find three things.

Difference 1 _____

Difference 2 _____

Difference 3 _____

How are they the same? Find three things.

Similarity 1 _____

Similarity 2 _____

Similarity 3 _____

Activity 2 (continued)

3 Talk about your ideas with your group.

4 Tell the class what differences your group has found.

Things that have never lived

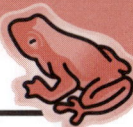

Activity 3

You will need: an area outside where you can explore things and a pen or pencil.

(a)  (b)  (c)  (d)  (e)
(f)  (g)  (h)  (i)  (j) 
(k)  (l)  (m)  (n)  (o) 
(p) 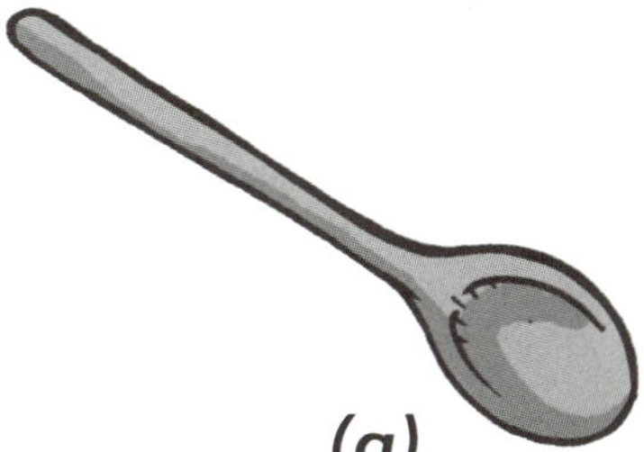 (q)  (r)  (s) 

7

Chapter 1: Plants

Activity 3 (continued)

1 In your class, get into small groups. Look at the pictures on page 7.

Sort the things into three groups:

> **Living things**
> **Things that were once alive**
> **Things that have never lived**

2

Living things

Things that were once alive

Things that have never lived

Write down the letters of the things in the correct group.

8

Things that have never lived

Activity 3 (continued) 📖

3 Go outside. Can you find one living thing and one non-living thing?

I found a living thing called a

_____

I found a non-living thing called a

_____

4 Back in class, put the things you have found in their groups.

Let the class see what you have done.

5 Look at the groups made by others in the class.

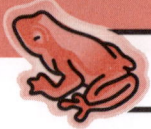

Plants and animals in their environments

Activity 4 📖 12

You will need: an area outside where you can explore and a pen or pencil.

In this activity you will explore a local environment in a small group.

1 Choose where the environment will be.
- It can be in a field.
- It can be on the sunny side of the school, or on the shady side of the school.
- It can be in a pond or on a beach.

These are just some examples.

We chose the _____

2 Choose what you will observe. Will it be:

the plants   animals   the soil   the water   the sunlight?

We chose the _____

Plants and animals in their environments

Activity 4 (continued)

3 Make a plan with your group to **explore** this local environment.

Plan

First we

Then we

Next we

4 Show the group's plan to your teacher.

Chapter 1: Plants

Activity 4 (continued) 📖

5 If you need to collect plants and animals, you must choose what to use. It could be:

a bag · a cardboard box · a jug

a bucket · a glass jar · your hands

Be careful when picking flowers or touching animals. Some plants and animals are dangerous. Your teacher will suggest good ones to choose.

Activity 4 (continued) 📖

6 **Think of how to use your senses.**

a What can you smell?

I can smell _____

_____

b What can you feel?

I can feel _____

_____

c What can you hear?

I can hear _____

_____

Activity 4 (continued) 📖

7 Choose how you will **record** what you observe and **collect**.

8 Work as a group. Share out the tasks.

   a Collect all the evidence you can about the plants and animals in the environment you have chosen. Use the box below to record it.

   b Decide how to show (display) what you have found out.

   c Add the name of the environment that you explored to your display.

Activity 4 (continued)

Chapter 1: Plants

Activity 4 (continued) 📖 15 – 📖 16

9 What will you see when you look at what the other groups found?

Tell your teacher what you think (predict) you will see.

I think I will see

10 Move round the class. What has everyone else found?

Is it what you predicted? Circle your answer.

Yes   No

Plants and animals in their environments

17

🔍 Look at the three environments shown in the pictures above.

What can you see that is different in each environment?

💬 Tell the class. Try to explain why some things are different.

What can you see that is the same in each environment?

💬 Tell the class. Try to explain the similarities.

The parts of a plant

18

stem   root   flower   leaf

Here is a picture of a plant.

Write the name of each part in the right box. Choose from the list of words.

Share your answers with the class.

The parts of a plant

Activity 5 📖19

You will need: some things to make models with and a pen or pencil.

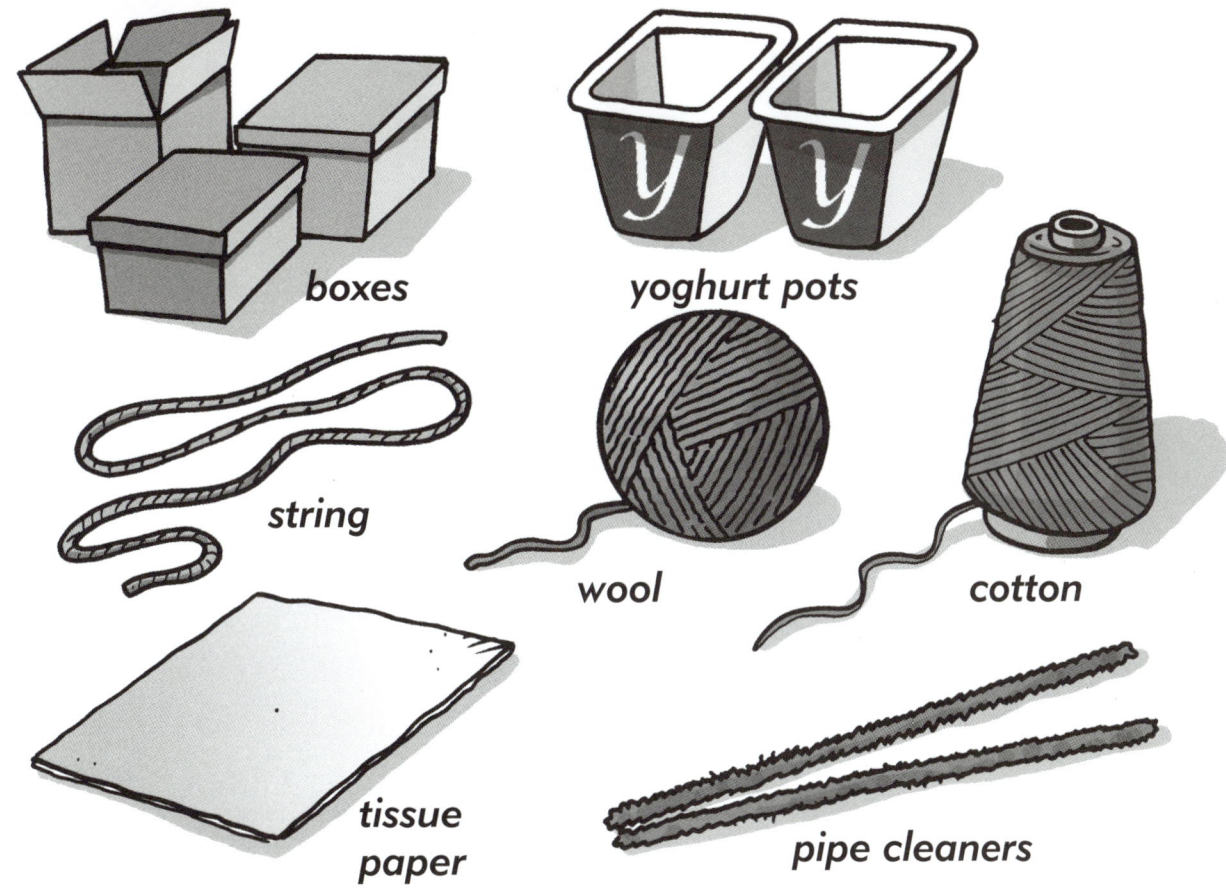

boxes · yoghurt pots · string · wool · cotton · tissue paper · pipe cleaners

🔍 **1** Go outside and find a small plant. Try to find one with flowers.

> Be careful when picking flowers as some can be poisonous. Your teacher will suggest good ones to choose.

2 Dig it up with care and take it to class.

Chapter 1: Plants

Activity 5 *(continued)* 📖 20

✏️ **3** Draw a picture of the plant.

Name each of the parts you can see.

Write the names on the drawing.

Activity 5 (continued) 📖20

4 Observe the plant closely.

  a Look at how the parts are fitted together.

  b Tell the class what you have observed.

  Use these words to complete the sentences below.

  > **roots   stem   leaves**

  The __ __ __ __ __ are in the soil.

  The __ __ __ __ is joined to the roots.

  The __ __ __ __ __ __ grow from the stem.

5 Make a model of a plant.

  Choose what you will use to make the stem, the roots, the leaves, and the flowers.

6 Display your model. Now look at those made by others.

The parts of a plant

What plants need to grow

Activity 6 📖24

You will need: a pen or pencil.

The class will do an investigation to find out:

'What are two of the things plants need to grow?'

1 Talk to your group about how you can find an answer to this question.

2 Share your ideas with the class. Write them down here:

What plants need to grow

Activity 6 (continued) 📖 24

3 **Plan the part of the investigation that your group will do.**

  a Plan how you will collect the evidence.

  b Plan how to record what happens.

Plan

First we

Then we

Next we

  c What do you think will happen? Tell your prediction to your teacher.

  I predict that _____

  _____

4 **Choose the things you need for the investigation and collect them.**

Chapter 1: Plants

Activity 6 (continued) 📖 25

💬 **5** Share your group's plan with the teacher.

Activity 6 (continued) 📖 26

 **6 Do your investigation and record what happens.**

   a Are the results the same as your prediction? Circle your answer.

   Yes   No

   Was it right or wrong?

   It was _____

   b Share the results with the class.

   c Listen and look as other groups report their results.

7 What is the answer to the question you investigated?

   a Tell the class what you think.

   I think the answer is _____

   _____

   _____

   b Try to explain why you think that.

   I think this because _____

   _____

   _____

Growing plants from seeds

Activity 7

You will need: a pot, some seeds, some soil or compost, water and a pen or pencil.

1 What will the seeds need to make them grow? Talk about it.

Growing plants from seeds

Activity 7 (continued) 📖 28

2 Share out the tasks in the group:

   a Collect the things you need to explore how seeds grow into plants.

   b Plant the seeds in pots.

   c Put a group name or number on the pot.

3 When the seeds have been planted and have what they need, put the pot in a good place.

   a Look at the pot each day.

   b Draw pictures of it. Record any changes you see.

   c Put the day's date on each drawing.

Chapter 1: Plants

Activity 7 (continued)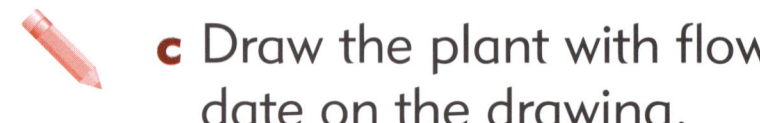

4 **Take care of the seedlings.**

a How can you stop them from falling over as they get bigger?

I can _____

b Try to keep your plants alive until the flowers open.

c Draw the plant with flowers. Put the day's date on the drawing.

5 Display your plants and your drawings.

28

Activity 7 (continued)

6 Look at the plants and drawings from other groups. Compare them with yours.

    a Are they different? Circle your answer.

    Yes   No

    b How are they different? _____

_____

7 Draw a plant that is different from yours.

    a Show it to the class.

Chapter 1: Plants

Activity 7 (continued) 30

b Tell the class what you can see.

I can see

_____

_____

8 Now you have finished, what have you found out about growing plants from seeds? Tell the class what you think.

I think seeds need _____ and

_____ to grow well.

Growing plants from seeds

Activity A

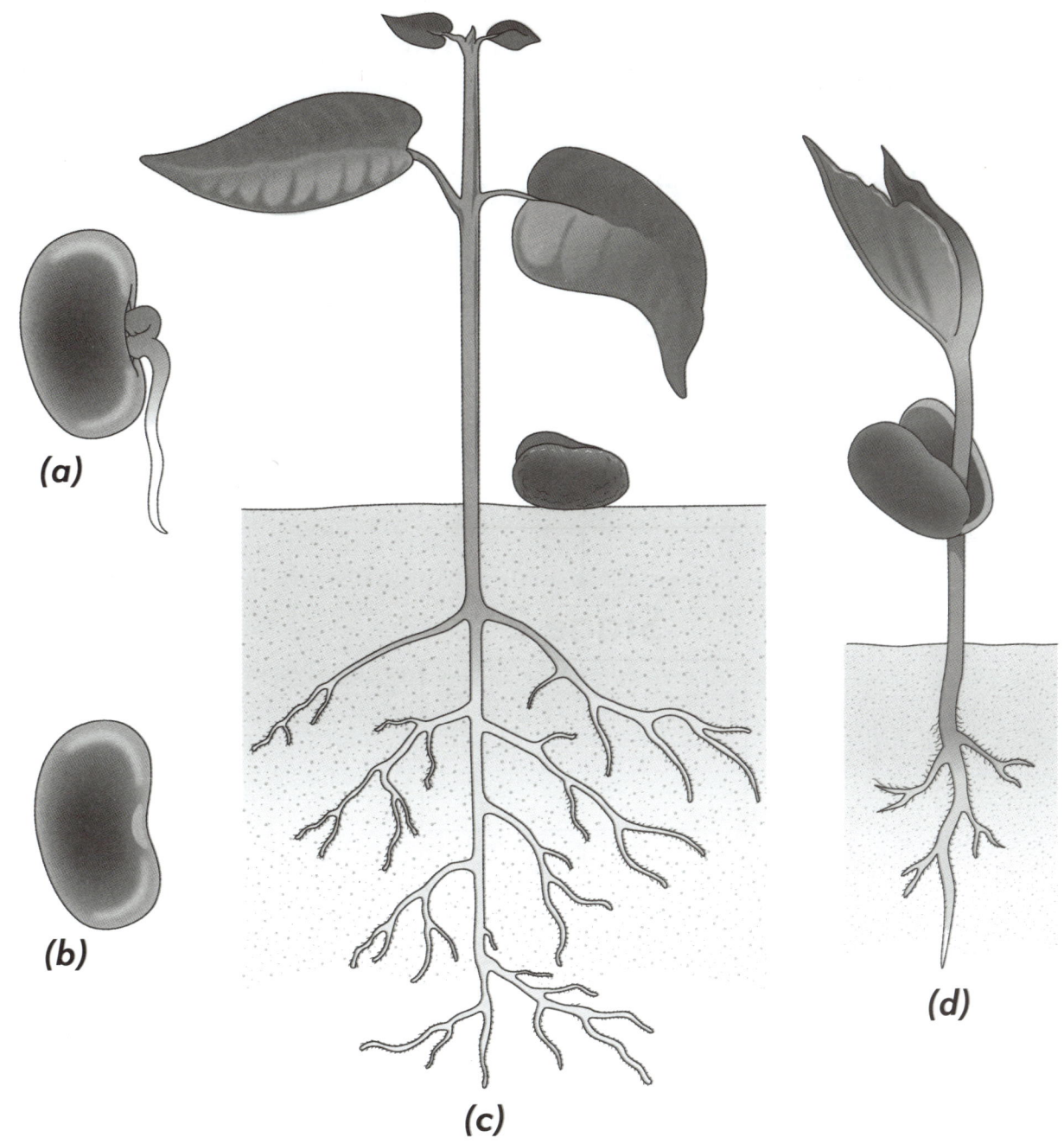

(a)
(b)
(c)
(d)

You will need: a pen or pencil.

1 The pictures show some stages (steps) of a seed growing.

They are mixed up.

31

Chapter 1: Plants

Activity A *(continued)*

2 Look at them and sort them into the right order.

3 What is the right order? Write the letters in the boxes below.

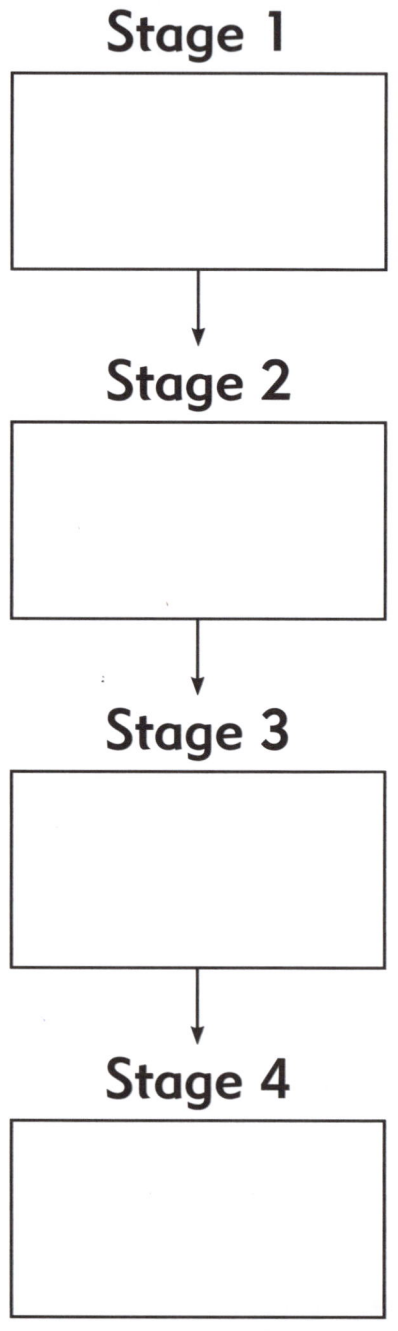

Growing plants from seeds

Activity B

You will need: three different leaves and a pen or pencil.

1 Collect three different leaves. Call them leaf A, leaf B and leaf C.

2 Draw them in the boxes below.

Leaf A

Leaf B

Leaf C

33

Chapter 1: Plants

Activity B (continued)

3 **Compare the leaves.**
Describe your different leaves.

Leaf A is

_____

Leaf B is

_____

Leaf C is

_____

4 **What is the same about the leaves?**

All the leaves

_____

_____

_____

5 **What is different about the leaves?**

_____

_____

_____

Chapter 2: Humans and animals

We are all different – and the same

Activity 1 📖 33

You will need: colouring pencils.

1 Draw a picture of yourself.

Chapter 2: Humans and animals

Activity 1 (continued)

2 Now draw a picture of another child in your class.

3 Display the two drawings side by side.

a Let the class look at them.

b How do you both look?

What is different about you?

What is the same about you?

We are all different – and the same

Activity 1 *(continued)* 📖 34

4 **Now do the same with drawings by another child in the class.**

The differences are

_____

_____

_____

The things that are the same are

_____

_____

_____

💬 **Talk about the drawings with the class.**

Tell the class your ideas about the differences and similarities.

_____

_____

Body parts

Activity 2 36

You will need: a pen or pencil.

Complete the sentences below.
Use these words to fill in the gaps:

> food   ears   speak   teeth   eat   eyes

1 We use our __ __ __ __ to hear sounds.

2 We use our mouths to __ __ __ and to __ __ __ __ __ and to breathe.

3 We use our __ __ __ __ to see.

4 We use our __ __ __ __ __ to bite and chew our __ __ __ __.

Body parts

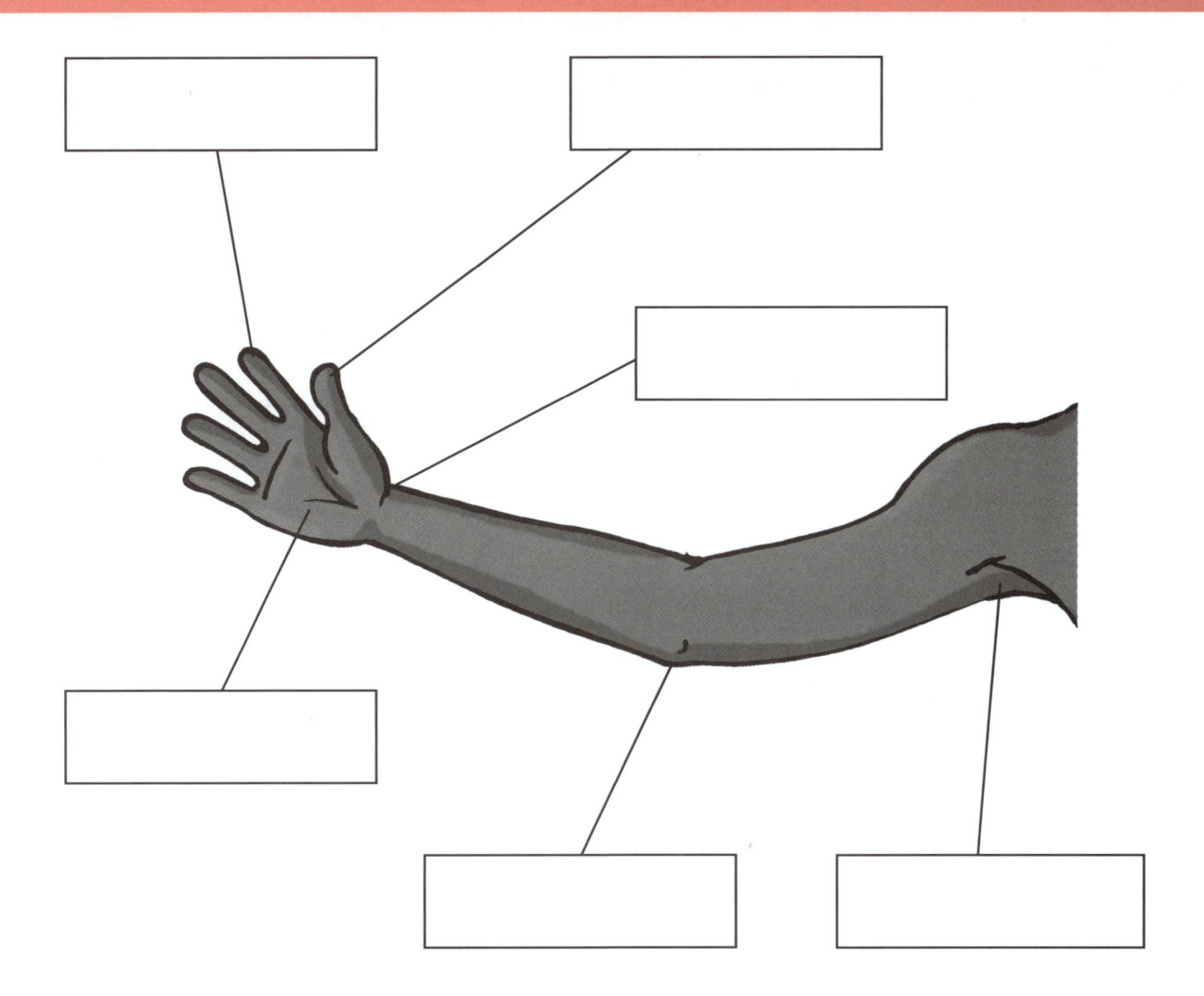

Activity 3 [37]

You will need: a pen or pencil.

1 Look at the picture of an arm.

   a Can you name all the parts of the arm?

   b Touch each part on your body as you name it.

2 Write the name of each part in the correct box. Choose from this list:

hand   finger   thumb   elbow   armpit   wrist

Chapter 2: Humans and animals

Activity 3 (continued) 📖 38 – 📖 39

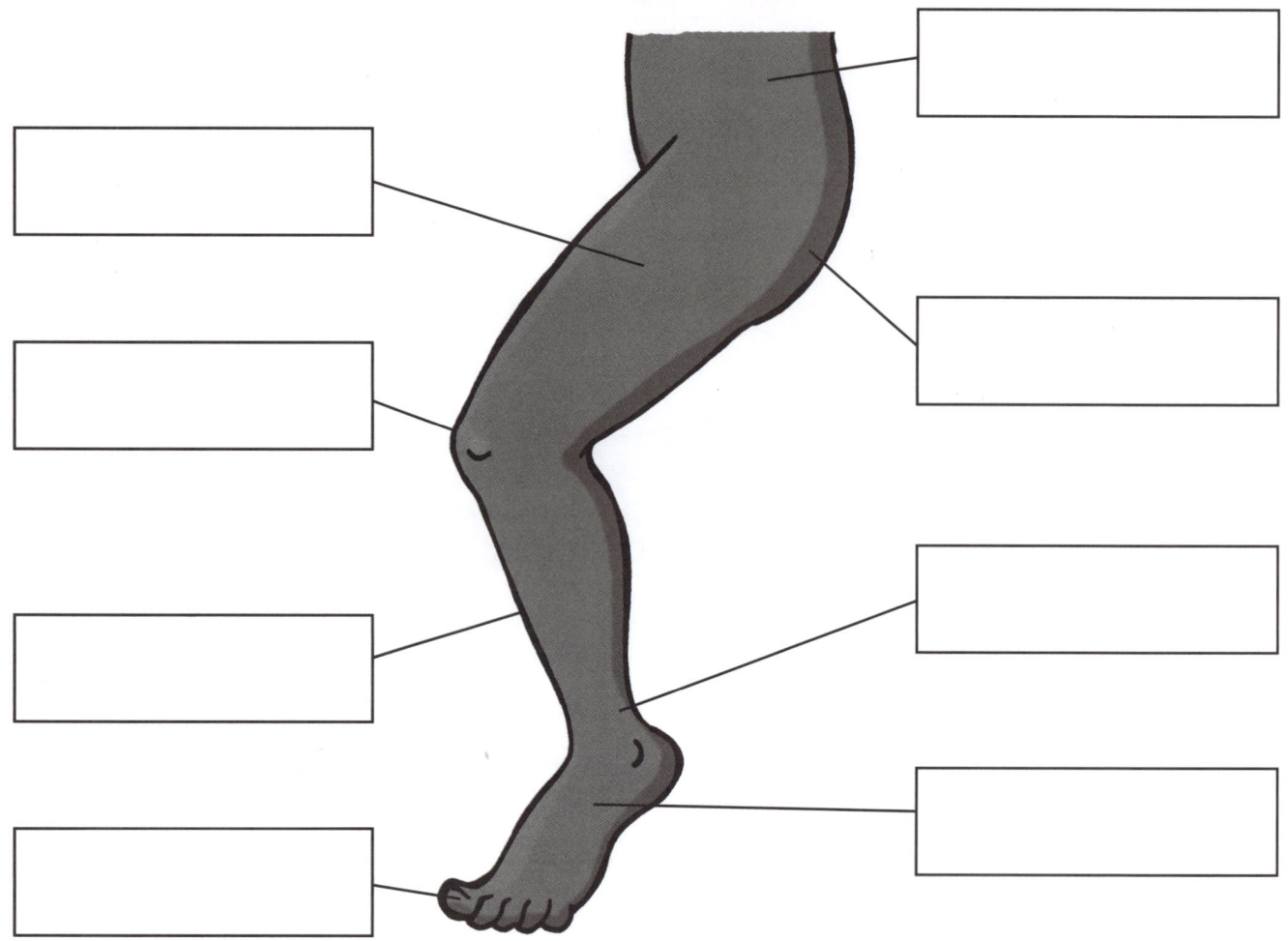

3 Look at this picture of a leg.
   a Can you name all the parts of the leg?
   b Touch each part on your body as you name it.

4 Write the name of each part in the correct box. Choose from this list:

ankle   foot   toe   knee   hip   thigh   shin   buttock

5 Do you know a song that can help us to remember the names of our body parts?

40

Body parts

Activity 4 📖 40

You will need: colouring pencils.

1 Complete this table.

Count how many of each body part you have. Write the numbers in the table.

The first row has been done for you.

We have two arms.

Body part	Number
arms	2
hands	
fingers	
eyes	
elbows	
ears	
nose	
mouth	
lips	
tongue	
legs	

Chapter 2: Humans and animals

Activity 4 (continued) 📖 41

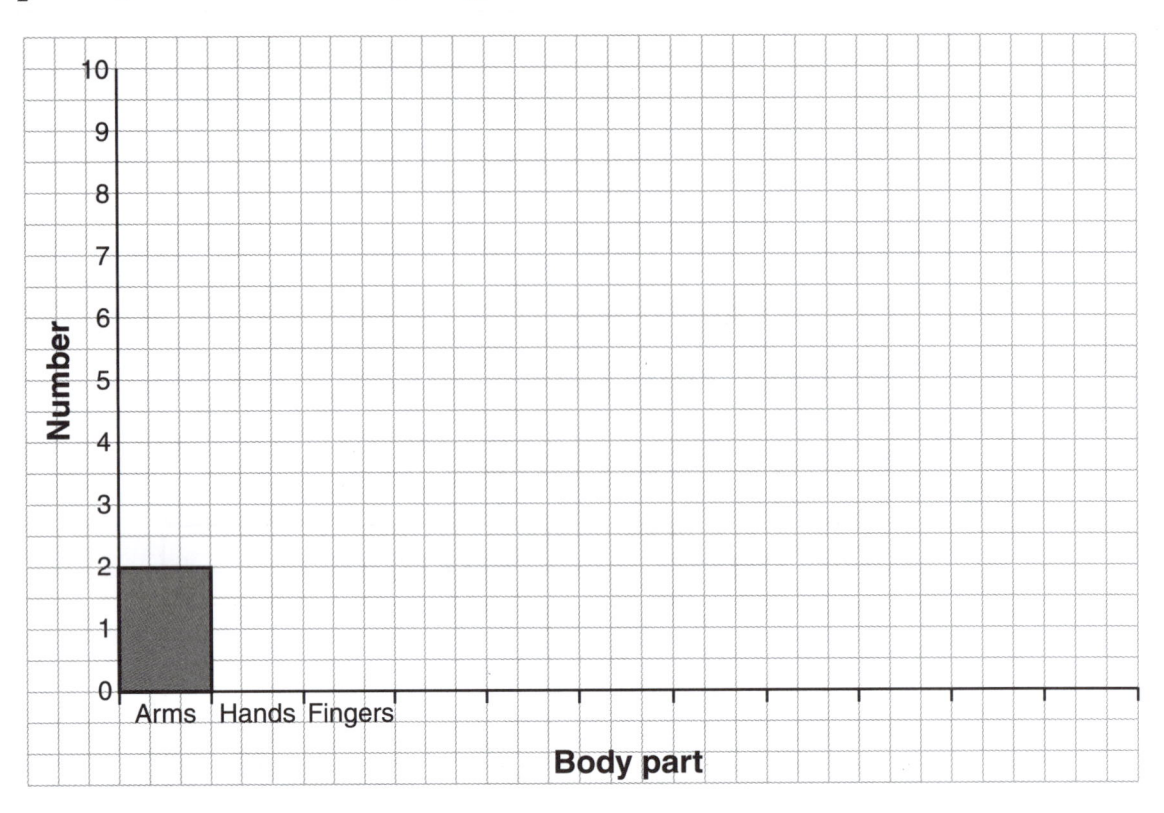

2 Complete this bar chart. Use your table to help you fill in the blocks for each body part correctly.

3 Show your chart to the class and say what it shows.

Health and diet

Activity 5 [42]

1 Look at these two pictures.
One meal is healthy.
One meal is unhealthy.

2 Talk to your group about the meals.
  a Which is the healthy meal?
  b Tell the class what you think.

  I think the healthy meal is _____

Chapter 2: Humans and animals

Activity 5 (continued) 📕 42

3 Explain why you think this is the healthy meal.

I think it is healthy because

_____

_____

Activity 6 📕 46

tap water

bottled water

pump water

well water

river water

pond water

stream water

🔍 **1** Look at these pictures of water from different places or sources.

44

Activity 6 (continued)

2 **Talk about them in your group.**

    a Choose a healthy water source.

    _____ is a healthy water source.

    b Choose an unhealthy water source.

    _____ is an unhealthy water source.

3 **Tell the class which you chose. Explain why you chose them.**

I chose them because

_____

_____

_____

_____

Health and diet

Senses

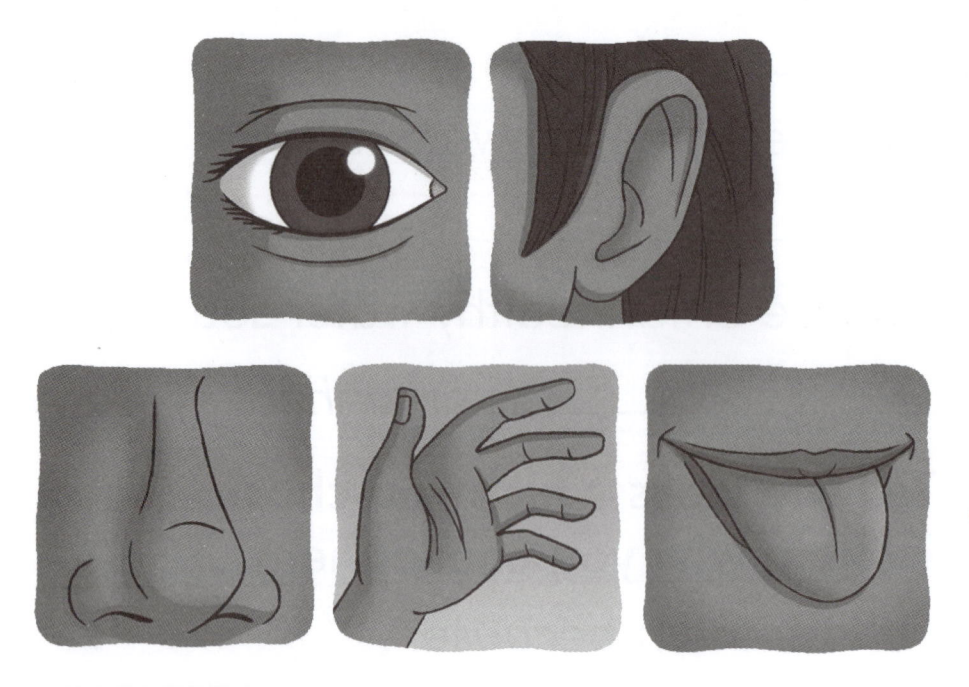

Sense organ	Sense
eyes	touch
ears	smell
nose	sight
tongue	hearing
skin	taste

Activity 7 49

You will need: colouring pencils.

Look at the table.

Draw lines to match each sense organ to its sense.
The first one has been done for you.

46

Senses

Activity 8 📖 50

You will need: an outdoor area where you can walk.

🔍 **1** Look at the children in the picture.
Which senses are they using?

They are using their senses of _____

_____

2 Go out on a nature walk.

3 Try to use all your senses.

⚠️ **WARNING:** do not put things in your mouth.

47

Activity 8 *(continued)* 📖 50

4 Back in class, tell others what your senses told you. Write it down here.

_____

_____

_____

_____

_____

_____

Our sense organs pick up the sounds, the sights, the smells, the tastes and the 'feel' of things.

Activity 9 📖 51

You will need: a scarf or piece of cloth, and a pen or pencil.

1 In your group, you will explore the way your senses help you.

Collect five different things. Choose things of different sizes, materials, feel, colour and smell. Some must make a sound.

Write them down here.

_____

_____

_____

2 Put the things under a cloth on a tray or table.

You will swap with another group so that you don't know what the objects are.

Then you will take it in turns to feel the objects under the cloth.

You will use your senses to tell the sound, the smell, the shape, the feel, the size and the material of each object. Work out what each object is, then name it.

Chapter 2: Humans and animals

Activity 9 (continued) 📖 51

3 Plan how you will record the results.

Plan

First we

Then we

Next we

Show the group's plan to the teacher.

Senses

Activity 9 (continued) 📖 52

✏️ **4** Record your answers. Write down the results for each turn.

👂 It makes a sound like

1	2	3	4	5

👃 It smells like

1	2	3	4	5

✋ It feels

1	2	3	4	5

Continued on next page

Chapter 2: Humans and animals

Activity 9 (continued) 📖 52

Its size is

1	2	3	4	5

The object is a

1	2	3	4	5

💬 **5** Compare the results.

What do each of your senses help you to do? Discuss each sense in turn.

Our ears help us to _____

Our eyes help us to _____

Our skin helps us to _____

Our noses help us to _____

💬 **6** Share your results with the class. Try to explain them.

Parents and children

(a)
(b)
(c)
(d)
(e)
(f)
(g)
(h)
(i)
(j)

Activity 10

You will need: a pen or pencil.

1 Look at the pictures of adults and their young.

2 Match the parents and their children.

Chapter 2: Humans and animals

Activity 10 *(continued)*

3 Write a list of the letters for each pair.

a goes with _____

b goes with _____

c goes with _____

d goes with _____

e goes with _____

4 Share your answers with the class.

Parents and children

(a)
(b)
(c)
(d)
(e)
(f)

Activity 11

You will need: a pen or pencil.

1 Look at these pictures of people of different ages.

2 Sort them into the right order – from the youngest to the oldest.

55

Chapter 2: Humans and animals

Activity 11 *(continued)* 📖 57

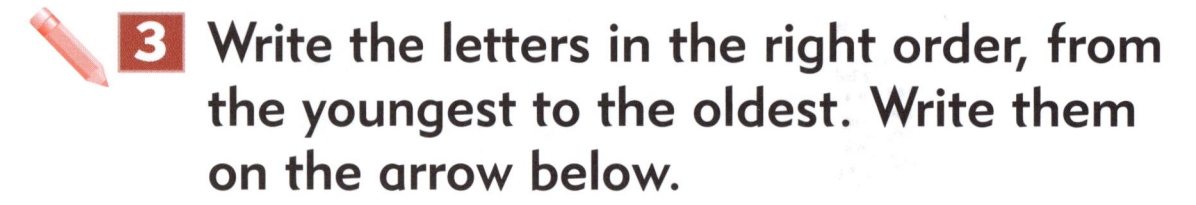

3 Write the letters in the right order, from the youngest to the oldest. Write them on the arrow below.

Youngest ⟵————————————⟶ Oldest

4 Share your answers with the class.

Parents and children

Activity C

**head   arm   leg   eye   lips   ear   hand   elbow
finger   knee   foot   shin**

You will need: a pen or pencil.

1 Look at the picture of the body.

   a Read the names of the body parts.

   b Match the names to the body parts.

   c Write the correct names in the boxes on the diagram.

57

Chapter 2: Humans and animals

Activity D

You will need: a pen or pencil.

1 Keep a food diary at home for three days.

  a Record all the things you eat and drink at home.

  Make a mark in the 'tally' column, like this | every time you eat or drink each type of food or drink.

	Food and drink	Tally	Totals
A	Egg		
B	Milk		
C	Fish		
D	Rice		
E	Bread		
F	Fruits		
G	Vegetables		
H			
I			
J			

  b If you eat or drink other things, write them in the blank spaces in the table.

2 Count up all the marks you've made in each row. Write the totals in the last column.

Parents and children

Activity D *(continued)*

3 Turn the data from the table into a bar chart. Draw and colour the right number of squares for each food.

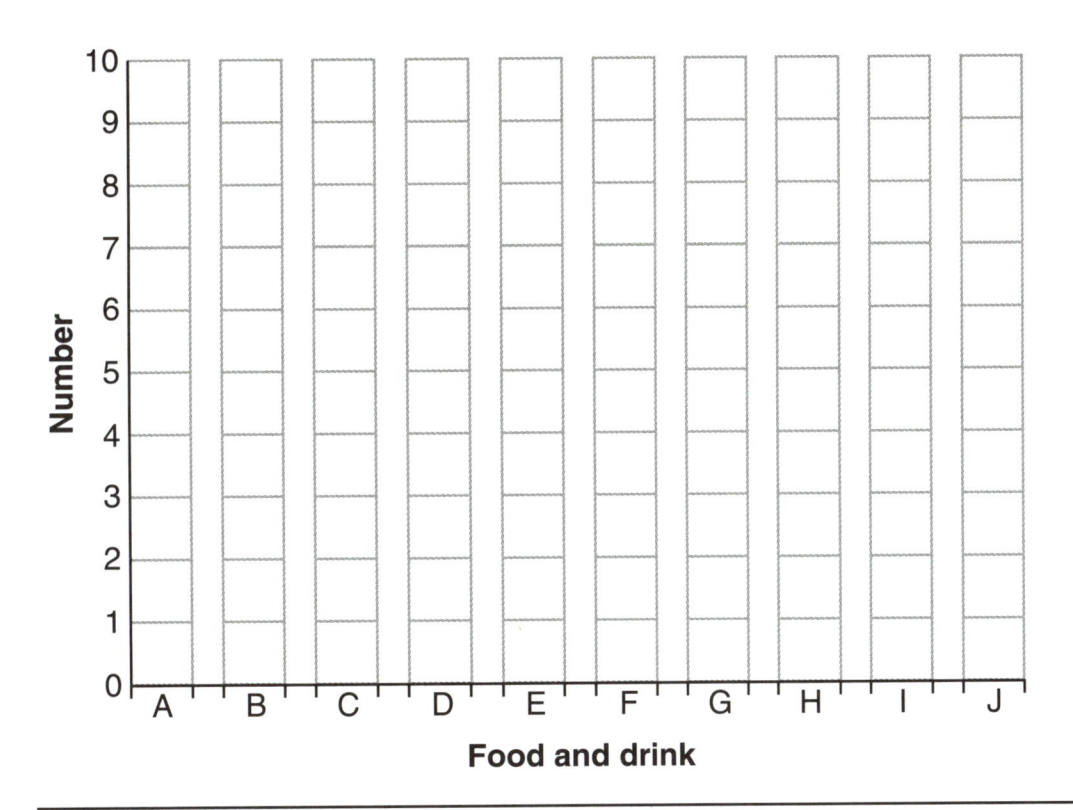

Number

10
9
8
7
6
5
4
3
2
1
0
A   B   C   D   E   F   G   H   I   J

Food and drink

Key

A = Egg   B = Milk   C = Fish   D = Rice   E = Bread

F = Fruits   G = Vegetables   H = _____   I = _____   J = _____

Chapter 3: Material properties

What are things made of?

Activity 1 │59│

You will need: a pen or pencil.

1 **Move around the classroom and find two different materials.**

Material one is _____

Material two is _____

2 **Compare the materials you have found.**

a What is different about each one?

_____

b What is the same about them?

_____

3 **Tell the class what you observed.**

a Show them what you collected.

b Tell them which senses you used to compare each material.

I used my _____

_____

What are things made of?

Activity 1 (continued) 📖59 – 📖60

4 **Go outside and find two other materials.**

Material three is _____

Material four is _____

5 **Compare the materials you have found.**

   a What is different about each one?

   _____

   b What is the same about them?

   _____

6 **Tell the class what you observed.**

   a Show them what you collected.

   b Tell them which senses you used to compare each material.

   I used my _____

   _____

   _____

Materials and their properties

Activity 2 📖62

You will need: a pen or pencil.

1 Look at the four things you collected in Activity 1. Touch them.

> red   hard   green   soft   yellow   smooth
> wet   rough   sticky   shiny   black   round
> sharp   blue

The words above are all **properties** (characteristics) of materials.

2 Choose words that describe your four materials. If the words you need are not in the list, then add your own words.

3 Draw your four materials and copy the words that describe them.

Materials and their properties

Activity 2 (continued) 📖 62 – 📖 63

4 Show the class what you have done.

5 Tell them which senses you used to identify the properties of your materials.

6 Play a game with your class.

Either:
Choose an object from your drawings. Name one property that it has.

Or:
Choose a property. Then choose an object that has that property.

Now get into your group and play the game again.

Naming materials

65

Everything is made of a material.

Each material has a name.

Some materials are put together in a group.

What group do water, milk and orange juice belong to?

They are all _____

What group do gold, iron and silver belong to?

They are all _____

Tell the class what you think.

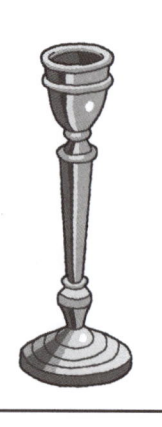

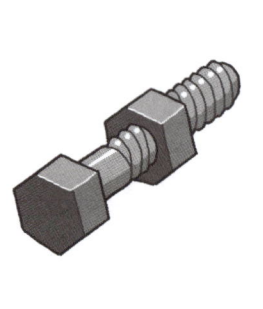

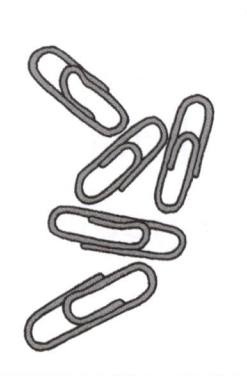

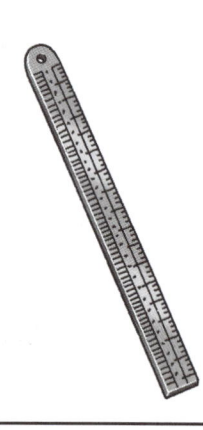

Naming materials

Activity 3 📖

You will need: a pen or pencil.

1 Look at the pictures on page 64.

2 Talk to your group about what each one is made of.

   a Identify (name) the materials.

   b Tell the class what your group thinks.

3 Now choose four of these materials: wood, metal, plastic, glass, fabric, rubber.

4 Move around the room.

   a Find two examples of each material.

Continued on next page

Chapter 3: Material properties

Activity 3 (continued)

b Collect them or make a drawing of each one.

5 Show the class what you have found.

6 Tell the class the names of the materials each object is made of.

Sorting materials

Activity 4 📖 68

You will need: a pen or pencil.

1 Collect three things that have at least one property that is the same.

I collected

_____

_____

2 Add them to your collection.

3 Mix up all the items.

   a Sort them into three groups. Take it in turns to do this.

   b Each time, choose a property for each group.

   c Don't tell the others which property you chose.

4 Look at the groups.
What is the property each group has?

Group A has the property of _____

Group B has the property of _____

Group C has the property of _____

Chapter 3: Material properties

Activity 4 (continued)

5 Tell the people you are working with what you think.

6 Look at the other groups collected by your classmates. Try to work out what property each of their groups has.

Group A has the property of _____

Group B has the property of _____

Group C has the property of _____

Sorting materials

Activity E

You will need: a pen or pencil.

1 Find three things.

One must be made of paper.
One must be made of metal.
One must be made of clay.

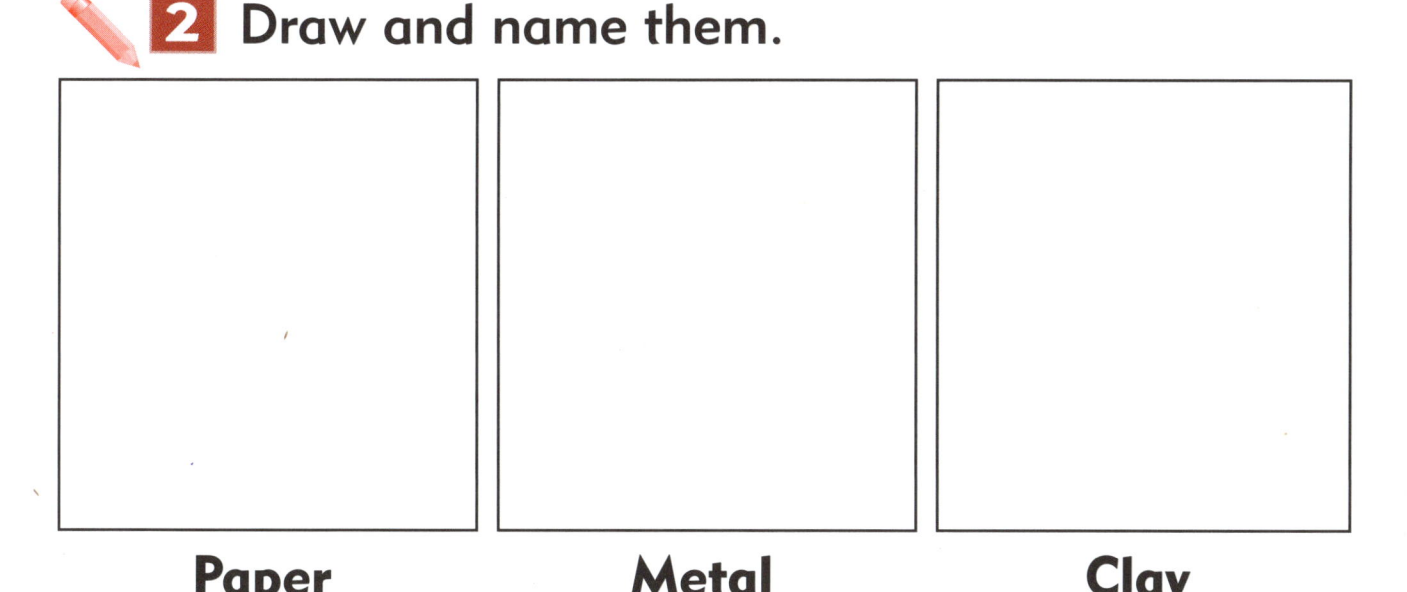

2 Draw and name them.

Paper	Metal	Clay
Name:	Name:	Name:

3 Finish these sentences. Add the names and a property of each one.

a Paper: The _____ is _____

b Metal: The _____ is _____

c Clay: The _____ is _____

Chapter 3: Material properties

Activity F

You will need: a pen or pencil.

A

B

C

Sorting materials

Activity F *(continued)*

1 Look at the three groups on the previous page. Work out why the things are in the groups.

2 Finish these sentences:

Group A are all _____

Group B are all _____

Group C are all _____

3 Find one more thing to add to each group.

4 Draw and name them below:

Group A	Group B	Group C
Name:	Name:	Name:

71

Chapter 4: Forces

Movement

Activity 1 [71]

You will need: a pen or pencil.

1 Plan with your group how you will explore the movement of *three* different things.

Like this:

Plan

First we

Then we

Next we

Movement

Activity 1 (continued) 📖 71

2 How can you describe the movement of your objects? Circle the correct answer.

    slide   roll   twist   bounce

3 Think about how you will record what you see.

4 What will the movements be like? Predict them.

I predict that object A will _____

_____

I predict that object B will _____

_____

I predict that object C will _____

_____

5 Show your plan to the teacher.

Chapter 4: Forces

Activity 1 (continued) 📖 72

✏️ **6** Carry out your plan for each thing. Record it here:

Movement

Activity 1 *(continued)* 📖 72

7 Talk about the movements with your group. Compare them.

   a What was the same?

   _____

   _____

   b What was different?

   _____

   _____

8 Compare what you saw happen with what you predicted. Were you right?

   _____

9 Share your results with the class.

Chapter 4: Forces

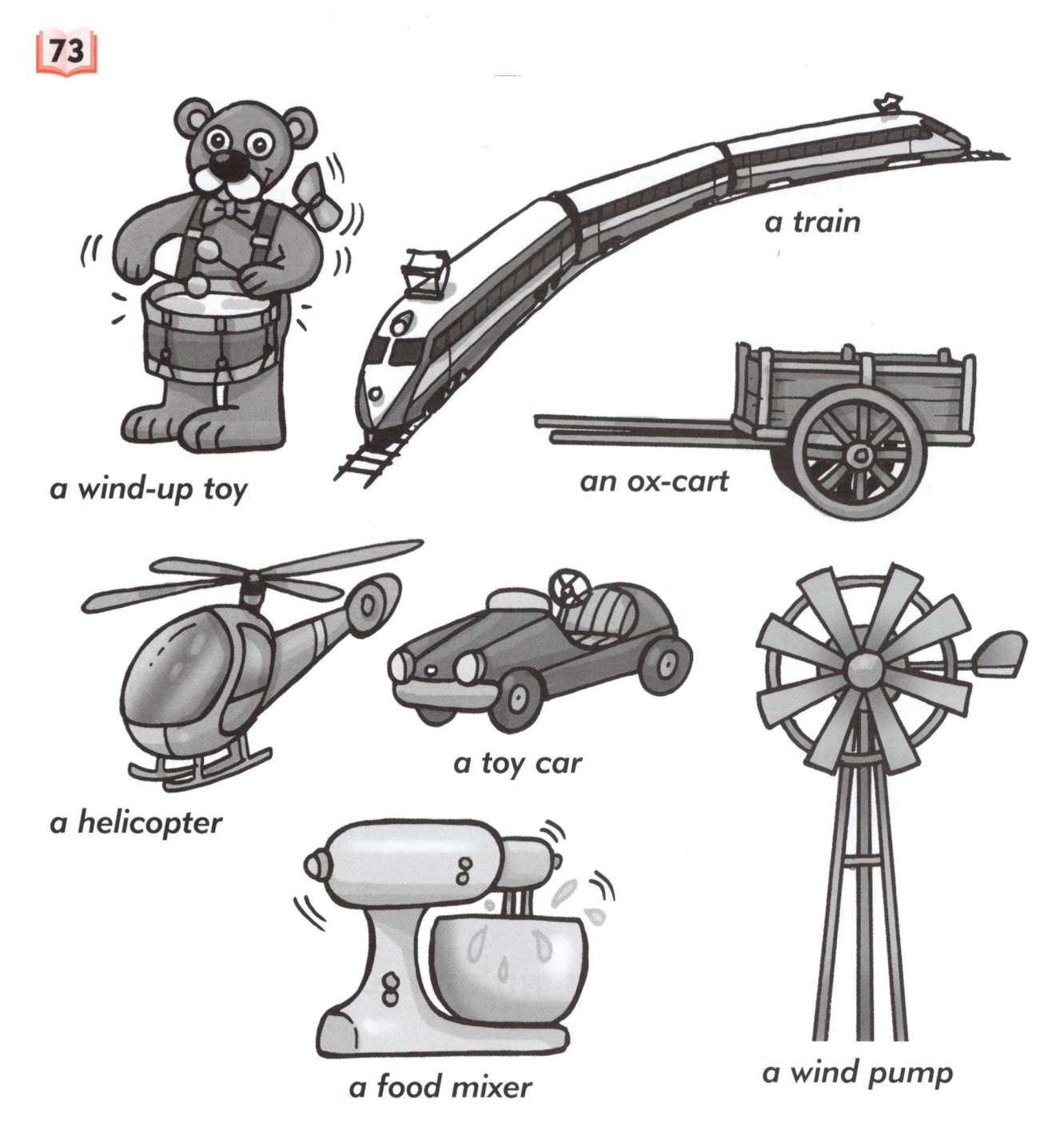

73

a wind-up toy

a train

an ox-cart

a helicopter

a toy car

a food mixer

a wind pump

Some toys can move.
Some have a motor to make them move.
Some can be wound up.
We can move some of them by hand.
We can push or pull them.

74

Complete the sentences below.
Here are the words you need:

> **wheels   wind   wings   kites   cars
> machines**

1 Movements can be made by

__ __ __ __ __ __ __ __ .

2 Bicycles, buses and __ __ __ __ move better because they have __ __ __ __ __ __ .

3 Planes have engines and __ __ __ __ __ .

4 Things can be moved by the __ __ __ __ .

5 The wind can move sails, flags and __ __ __ __ __ .

Pushes and pulls

75

Things move when they are pushed or pulled.

A push is a force.

A pull is a force.

(a)

(b)

(c)

(d)

🔍 Look at these pictures. They show forces making things move.

💬 Talk with your class about them. What can you see?

Pushes and pulls

75 – 76

Work out which are pulls and which are pushes.

✏️ Write down your answers, using the letters (a) to (d).

The pushes are _____

The pulls are _____

Share your answers with the class.
Explain your answers.

_____

_____

_____

_____

_____

Chapter 4: Forces

Activity 2 📖 76

You will need: a pen or pencil.

1 Use your hands to move four things in the room.

2 Identify the force each time. Is it a push or a pull that you use?

3 Record what you push or pull.

Write down or draw what you do.
Tick whether it is a 'Push' or 'Pull'.

Action	Pull	Push
1 Open door		
2		
3		
4		

4 Share your results with the class.

Changes in movement

Activity 3 📖 78

You will need: a ball or toy car.

1 Put a ball or toy car on the floor.

2 Make it speed up.

🔍 **3** How did you do it?

_____

_____

_____

Tell the class.

4 Make the ball or toy move. Slow it down.

Chapter 4: Forces

Activity 3 (continued) 📖

5 How did you do it?

_____

_____

_____

Tell the class.

6 Make the ball or toy move. Change its direction.

7 How did you do it?

_____

_____

_____

Tell the class.

Changes in movement

Activity G

You will need: items to push or pull and a pen or pencil.

1 Use your hands to move five different things.

2 Name the force you use each time.
Is it a pull or a push?

3 Record the names of the things you moved.
 a Write them in this table.
 b Tick whether you pushed them or pulled them.

Things I moved	Pull (✓)	Push (✓)

Chapter 4: Forces

Activity H

Force _____

Force _____

Force pull

Force _____

Force _____

Force _____

You will need: a pen or a pencil.

1 Look at the pictures.
A force is being used in each one.

2 Work out which force is a pull and which force is a push.

3 Write **pull** or **push** under each picture.

4 Draw arrows on the pictures to show which way the push or pull goes. One has been done for you.

Chapter 5: Sound

Sources of sound

Activity 1 📖 84

You will need: an area outside and a pen or pencil.

1 Go outside. Find a good place to sit still and be as quiet as you can.

2 Be very still. What can you hear?

I can hear

_____

_____

_____

3 Do you know what is making each sound?

4 When you have heard four different sounds, tell your teacher.

5 Back in class, identify the sounds you heard.

Sound A is _____

Sound B is _____

Sound C is _____

Sound D is _____

86

Complete the sentences below.

Here are the words you need:

> thunder wind crying sea frogs clapping
> dogs talking birds singing

a Some sounds are natural, like the s __ __ , the w __ __ d and t __ u __ d __ r.

b Some sounds are made by animals, like b __ r __ s, f __ o __ s and d __ g __ .

c We can make sounds, like s __ n __ i __ g, c __ a __ p __ n __ , t __ l __ i __ g and c __ y __ n __ .

Sound and distance

Activity 2 📖 87

You will need: an area outside and a pen or pencil.

1. Go outside and stand in a circle. Face outwards.

2. Listen to the sound made by the teacher.

3. Slowly move away from the teacher. Listen for the sound.

Activity 2 (continued) 📖

4 Stop and listen. What do you notice? Try to explain it.

_____

_____

💬 **5** Predict what will happen if you go further away from the source of the sound. Tell the class.

I predict that

_____

_____

6 Slowly move further away. Listen for the sound.

7 Was your prediction correct? Circle your answer.

Yes   No

💬 **8** Try to explain what you have observed.

_____

_____

_____

Hearing sounds

Activity 3 📖 90

You will need: a pen or pencil.

1 Plan with your group to **explore** how you hear sounds.

2 Choose the things you need. Collect them.

Chapter 5: Sound

Activity 3 (continued) 90 – 91

3 Plan what you will do and how you will do it.

Like this:

Plan

First we

Then we

Next we

4 Tell your teacher what you plan to do.
Tell you teacher what you think will happen.

I predict that _____

_____

Hearing sounds

Activity 3 (continued) 91 – 92

5 Do your exploration.

   a Make observations.

   b What did you find out? Record the results.

6 Compare your results with your prediction. Is this what you thought would happen? Circle your answer.

Yes   No

7 Tell the class what you think about what you found out.

_____

_____

_____

Chapter 5: Sound

Activity 1

You will need: a pen or pencil.

A child went out to listen to sounds.

Here is the table of data they collected.

Source	Number of times sound heard	Totals
Birds	‖‖‖‖ ‖‖‖‖ ‖	
Cars	‖‖‖‖ ‖‖‖‖	
Babies	‖‖	
Dogs	‖‖‖‖	
Planes	‖	
Trains	0	

1 Add up the data for each source.

Write the numbers in the Totals column.

‖‖‖‖ is 5 times.

Hearing sounds

Activity 1 (continued)

2 Use the data to draw a bar chart. Colour it.

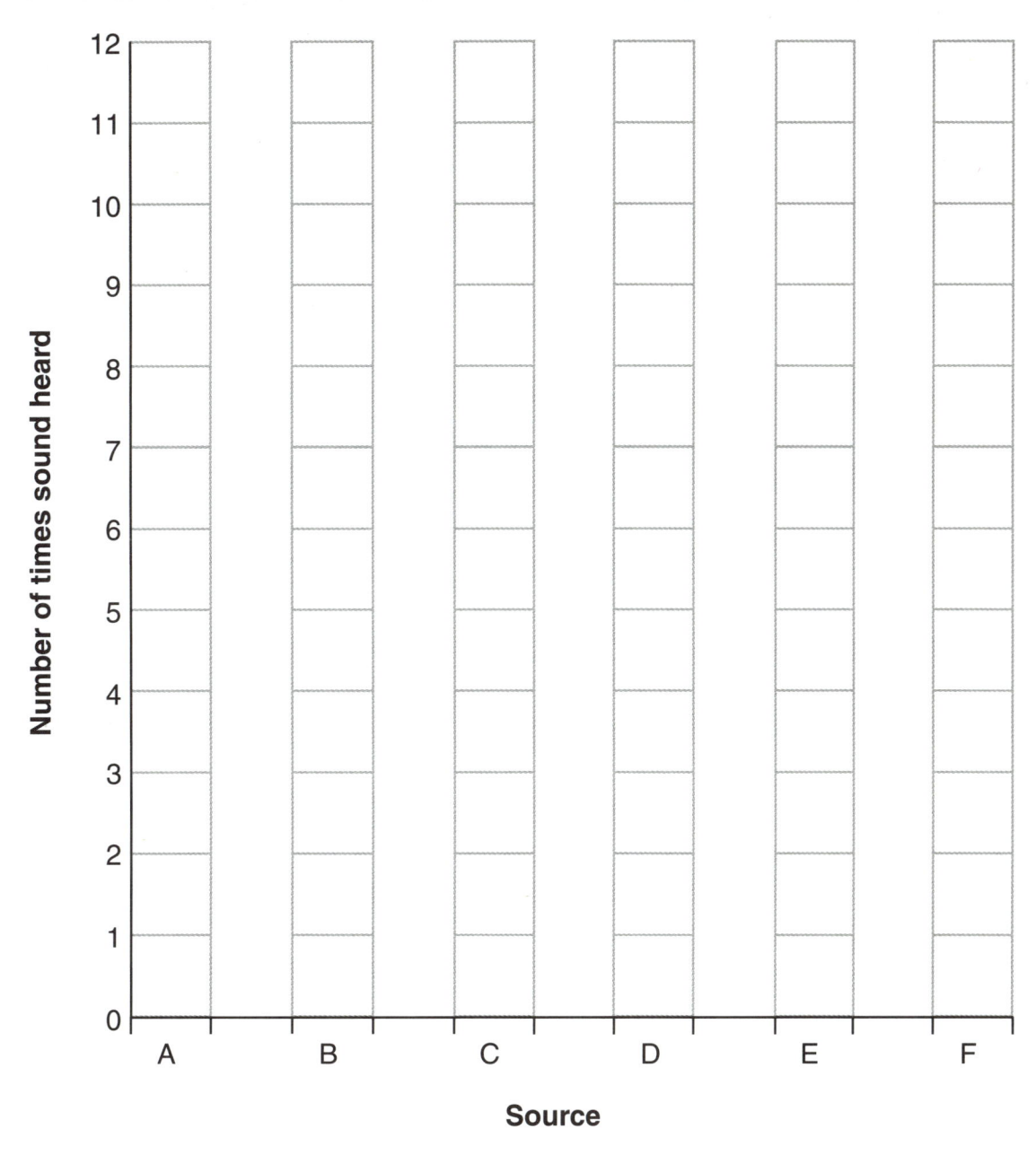

Number of times sound heard (y-axis: 0–12)

Source (x-axis: A, B, C, D, E, F)

Key

A = Birds   B = Cars   C = Babies

D = Dogs   E = Planes   F = Trains

Chapter 5: Sound

Activity J

You will need: a pen or pencil.

Complete the sentences below.
Here are the words you need:

> ears   source   natural   sense   far   faint
> travels   source   hear   wind

1 If the source of a sound is too __ __ __ away we cannot __ __ __ __ it.

2 The sound will be too __ __ __ __ __.

3 We use our __ __ __ __ __ of hearing to hear sounds.

4 The sound goes into our __ __ __ __.

5 Birds and the __ __ __ __ make __ __ __ __ __ __ __ sounds.

6 Sound __ __ __ __ __ __ __ away from its __ __ __ __ __ __.